THE CODEX

ANSWER KEY OF THE ANCIENT ASTRONAUT THEORY

PART I

by

James M. Gonzales

Word total
10502

THE CODEX

ANSWER KEY OF THE ANCIENT ASTRONAUT THEORY

PART I

by

James M. Gonzales

Contents

	INTRODUCTION	4
1.	Universal Life?	8
2.	Earths Space Travel	15
3.	The Worlds Ignored	23
4.	God of Man or Extraterrestrial?	31
5.	Ancient Histories Gods from Above	38
	End of Part I	44

*To be continued...

Introduction

    This book is for the brave. Written with daring and motivational intent. Not to question anyone's own theory of psychology, history, archeology, or religion. We as a people have been asking for more proof and a deeper clarification referencing ancient astronaut theory. In my opinion ancient astronaut theory is a very respectable informational resource. It is at least one perspective that I can confidently say I agree with. Taking notice to the opposing scientifically educated opinions. Claiming pseudo-science these are no more than closed minds with negative reference to protect their private interest on research or another project affiliation. In addition to proving these dictators wrong. History's facts of our so recent but ancient past has inspired me on a much broader scale.

    This entire book series will include bias and non-bias examples to highlight the unexplainable among the traditional situation in our society's normal functionality. Not only to emphasize our own evolution but to encompass the oblivious implementational behaviors. Alluding our perception concluding a civilizational purpose that's created by us, for them, and beyond life as we have known today. Introducing ground breaking informational theory to dilute the suppression. "Utilizing my (MBDPT) Multibody dynamic perspective theory". Known as: (The Answer Key) created by James Michael Gonzales: Author of: The Codex Series, hinting at collaboration with Ancient Astronaut Theorist, and calling the subject inspiration. Turning

(MBDPT) into the root of The Codex Series making it an official scientific method called: Multibody dynamic perspective theory or (MBDPT).

Contradicting the observation of educated simple minds. Emphasizing the hypocrisy and absolution behavioristics within a governed system. The new (MBDPT) scientific method: multibody dynamic perspective theory brings depth to inspire your minds curiosity. To me ancient astronaut theory is not a religion or movement but a very respectable resource of information for the bold. That I am thankful to have studied and participated amongst.

This book concludes psychological influence showing you the curve that bends your perspective. In turn its effecting the light from which you can see. Beholden amongst the darkest place of your mind's resource-fulness. Just as archeology and its physical evidence brings us truth of existence. Religions test of time ties our ancient history and archeology together giving us our own psychological impression. You can spend a lifetime studying before you reach the full spectrum of our reality among planet earth. "Scratching the surface of circumference around our existence is The Codex". By James Michael Gonzales

You're not here by accident. The Codex introduces a real level playing field of psychological ideology. No one human being can put the puzzle together without seeing the pieces from my new scientific method: multibody dynamic perspective theory (MBDPT). Emphasized from the foundational pillars of our doctoring existence. By utilizing psychology, history, archeology, and religion to produce a real effecting transitional thought on the informational understanding from within your mental perspective. Therefore, you shall see with a brighter light accessible from within your own mind. Realizing time tells us we have inconsistencies among our temple of creation. My (MBDPT) theory derives from psychology, history, archeology, and religion intergraded. Working or being utilized as one system for hypotheses emphasizing these categories as a compass or crucial vantage point. Today we use to gauge clarification of our existence of humanity. Making it a newly independent scientific method. Not yet accepted by governed science. Who makes huge leaps and reminds us of opinions that have one sided perspective? Then tells us what we can and cannot accomplish. This book reminds educated traditional minds by highlighting a truth that cannot be overlooked.

Ancient astronaut theory is a major conversation piece among today society. What is the meaning of time? Can you reference the temple of our creation? These are the type of questions this book will shine a new light on. Increasing readers depth by utilizing my multibody dynamic perspective theory (MBDPT).

Allah, our God, our Lord, and savior Jesus Christ. What about before Christ the Gods that came down from the stars above. Theism is your choice to choose what suits your fundamental beliefs. This book provides tons of information with bias and non-bias depth. This reader's choice book and series is full of great work by an independent twenty first century autistic polymath. The Codex has been written to those who have really asked themselves, why are we here? I attend mass and believe in theism. My perspective is not fulfilled because religion is only one piece of the puzzle and that's great. The Codex is for the greater good of human civilizations curiosity. Multibody dynamic perspective theory (MBDPT) derives from psychology, history, archeology, religion, with bias, and non-bias depth. To fill the void of the seeking truth you all require. I am giving you the instruction to find what you seek.

I want to thank all my family for the strength to write and share my work. A very special thanks to all Ancient Astronaut Theorist around the world and anyone else who has a positive opinion on the matter. Enjoy. – James M. Gonzales

CHAPTER ONE

Universal Life?

This is not considered a traditional conversation amongst todays society. After the initial shock, I know for a fact many others wonder about the same subject. Realizing how many different species of life reside here in our own planet. Just shy of ten million, we are not alone here on earth. Visualizing how small we are compared to our own planet, but remember Earth is a fraction in size compared to other planets even within our own solar system.

Most don't agree that the Earth is life's training ground, but I say yes. Earth is not classified as a dwarf planet and sure maybe a bigger more suitable planet for humans would be more realistic. All life is my clarification. Earth is the harbor of universal life. We have not even discovered all the species within our own planet. Our planet is covered by more than seventy percent water today. The most famous observation is that we need water to sustain human life. As we all know humans cannot survive in water. How can Earth be our natural habitat? Why are we on a planet covered in water? It's because Earth is the home to universal life not just us.

As traditional text suggests big bang evolution, research shows actual evidence of advanced knowledge during ancient times. Science includes numerous information twisting the perspective through out our civilization, telling us what we can, and cannot consider to be realistic. We need to start thinking on a broader perspective, so vast that we categorize our information gathering based on what it is we are trying to calculate. I love science, but it is almost directly blocking people from the pieces they need to break free from a single perspective of perception that effects our progression as a civilization.

We do not need to wait around for bacterial organisms to evolve on Mars. Secrets on secrets makes education controlled, information restricted, and manipulates our understanding? The difference is now among the current time. Earth is alive and well, the need to utilize all our information to capture universal life therefore it has not been confirmed yet. Consider a filter of informational knowledge that cannot be accepted endless its approved. Several questions about species that maybe they have never heard of but if you see them it makes no matter. As an example, what if a government scientist or representative sees them? Traditionally then we might have a new species documented or another classified file.

Universal life has been publicly announced several times, but the response was not what we expected? Somehow, I have not been surprised by the community response. It would almost seem purposely down played for so many years? The reality of it is that there are too many species to keep up with. There are signs of new life and evolved species developing right here on our planet Earth. It is more than intriguing, it has become alarming.

Our government has transitioned into a sixth military branch under executive order of the President in two thousand and eighteen, with urgency to print more funds just to ensure we have a Space Force Branch. Direct indication that we are running out of ground space, ourselves as we to as a civilization are expanding. Third world countries on Earth are implementing house hold occupancies regulations. Different nations are at boarder wars over population inflation. NASA is on the brink of terraforming Mars, but we thought that was only in the comic books with teleportation.

    Universal life is thriving right here amongst our own planet and has been since before our existence on Earth. The theme behind misleading representation is very simple to understand, but we can't blame our government entirely for doing what they are elected to do. The media is a front runner of the misleading the trend setting department. I just hope that the smartest people in the world don't fall victim to some of the most common errors while processing searches for the evolution of life in space. Chasing bacterial organisms in outer space is the biggest waste of money and time. Hints to why this brings such a huge uproar and negative response to this incredible subject. Here right under our noses are thousands of species that carry endless amounts of new research opportunities to bring us data that can help us advance hundreds of times faster.

    Take a moment to pick a species from within the Earths water alone, you can easily think of multiple. Here would be a fun fact to think of when referencing universal life. There are three hundred different types of Octopoda, Cephalopod Octopus in our ocean that we know of. They categorize the Octopoda, Octopus along with the traditional leach from a common body of water known as a lake or dam. I didn't just pick the Octopoda for no reason, this is obviously a clear example of living organism evolution happening right here on our planet. This species has been in our water file for over two hundred years confidently. One of the newest types of Octopoda, Mimic Octopus that impersonates other species and its surroundings, that's an award-winning defense mechanism. This new species was just discovered by accident within the past twenty years. To see how dramatic the transformations can be ground breaking. An this is one only one that we are referencing as an example to highlight the magnitude. Every report we get from NASA, and governmental officials is completely washed over, they part take in cover up without a doubt but again that's their job.

Being a United States citizen and living in North America. I have spent my time in traditional public education. Keep this in mind I wouldn't say I am the smartest person, but an average intellectual quotient score would be fair to suggest. During my child hood years, I grew up close to the Pacific Ocean our within driving range. Therefore, I used Octopoda, Octopus species as a specific example earlier to bring fruition that I never once believed in this species as a child. Until I was twelve years of age Octopoda, Octopus were a science fiction species completely not even real. I truly have heard of them in stories, movies, and even seen them depicted in places as art growing up. Never once in my mind did, I ever believe they were real. I felt this would be such a good example to compare not only the ancient astronaut theory amongst todays traditional minds, but also including the concept of universal life right here as it is unfolding on Earth.

Huge excitement diving in with examples, expressing different scenarios, and possible outcomes here on our home planet. Seventy percent water, only one species of a family of over three hundred. Here's a scenario to process humans expand there living space with the number of residential occupancies within their residence. Do you think that being on land for us it's different or changes that perspective at all for sea life? When I go swimming putting my self within reasonable space from another swimmer makes me more comfortable. I suppose if I lived down there space would be nice. On land surface we have evolving life even in the most minimal space that bigger species do not inhabit. I would suppose temperature makes a difference. On land different animals migrate, but water species do the same. Truly comparing and contrasting could fill up data sheets and over load hard drive around the world. In the end earths universal life cycle is evolving from species globally with attributes that are traditionally remarkable.

Indication of bacteria that has survived the horrific entrance through our atmosphere breaking through Van Allen belt aboard one of our craft returning from deep space. This is a poor example of telling us bacteria increases in outer space and can successfully pass in and out of our atmosphere. Government, using their own descriptions of life creation no matter how confusing and when simplified suggest the same universal life concept better known as the vacuum. Earth is the light at the end of the tunnel of space, no matter how you dress it up.

All universal life will reside here on earth. We are the only species looking to escape our own planet in case of threat or disaster. Due to the bounce effect, the vacuum of space is bringing any and all quarks, or even light particles directly to us. We aren't searching for life in space, just making space travel safe. Ensuring habitability without much clean up, or proof of past life. Just in case of return our government has reserve agriculture seed to reproduce Earth sustainable crops underground. Some of the wealthiest people on Earth already purchased their room and boarding pass tickets. Mars is the second home of human life's destination, but to be honest it might just be a retuned welcoming for us due to the evolution of universal life.

Land, habitat of human life currently residing amongst a planet that is covered with over seventy percent water. Population increasingly fluctuates dramatically, our air becomes intolerable, and the soil is drenched. The water levels rise, as the species within the waters grow. It would seem we are not alone and especially not the majority. Earth is thriving with new life and evolved species. Our preparation is abandonment of evolution and terraforming a new home. Restrictions vary, pro universal life, or epidemic? Earth the home of universal life not just us. Changing the perspective of habit will eliminate all the bias and non-bias bickering. Clear the status quo, give your mind a chance to see from outside the traditional concept.

We never needed to search for extraterrestrial universal life among the stars, its right here amongst us here and now. History, archeology, and religion has many documentations in the form of text, depictions, and hieroglyphics to advise us. The system of government has transformed societies exchange of goods system to a global priority. Minimizing, if not eliminating the conceptual beauty of life away. Putting precedence amongst every citizen within our civilization to part take in a rat race for an invisible notion of acceptance or regulated forms freedom.

Not enough, while empty homes are falling apart from inside out. Most working mothers and fathers don't even know their kids, work to build a relationship, or rebuild a destroyed one. At what cost with little to no time to spare. The child care and education system raise our children. Frustrated kids with guns, most just want a hug, or a normal home, tired of being beat up, and trying to fit in. Sick of waiting for free lunch, can't stand sitting alone, going back to an empty home and just want something positive to look forward for. Take notice citizens committing mass genocide, finger pointing, blame and then it's back to the same. Embracing a confused notion with hopes and prayers. Thinking the answers will be given for us and sweeping past the fact of solution. Dismissing our most unique quality given from above. Broken family's over ninety percent due to infidelity and money. Work, work, work, and what about these kids?

Missing truth, the power of intelligence from above to make a difference for good. Looking around, all I am seeing is human genetic organisms being manipulated by governmental actions. Unfortunately, the good book is on the backburner. Todays clock in and out on hourly rate method is the world's worst way of gaining income but most popular practice. Materialistic items and behavior have superseded our commandment of today's life existence? Unless, tragedy takes place on a massive scale but even then, today it's so common. If there is no suicide or glory to Allah at the end the clicker gets passed, that's the end of that. The real message wasn't even seen or noticed. Support for one another has dissipated, the new life is selfish. Realizing how special life really is, has become a huge task. See, hear, smell, taste, and touch. Earth, fire, water, air, and life. We currently live on earth where universal life should take precedence because it only evolves within our world to this day.

CHAPTER TWO

Earths Space Travel

Bright stars just passed the clear skies. Seems only so right from Earth but the hidden unknown darkness of space is the greatest challenge. Since the beginning of our existence we search for the heavens, creators, or the Gods from above. Hoping and wishing that the bright shining stars will ultimately be our savior in the future. The large task of the unknown will be our greatest accomplishment.

At the close of war intelligence sparked a rise of a new era. To the Moon, we must achieve new heights. Space travel takes us back to such a huge moment in history. Post disaster and tragedy looming. Why at that time after so much devastation? What did we find, or who did we find? Two different nations at a race for Technological advancement. The fruits of war. That's just it we could have risen in any category for the taking. Something that was so public and echoed through the ages after announced, today does not seem to have the same impact. Hundreds of thousands of reports civilian and military notice something else is among us.

After we accomplished the goal to reach space successfully, sudden withdraw came. A public informational cast of this moment loses its feed. Cutting communication, suppression, and systemic government takes precedence. A post press conference with returning astronauts from the first landing on the moon tells a different story then a quest of success. Today, we know that there was visualization of unexplainable objects witnessed multiple times during this journey. As we have ventured forward with so many travels into space none of these ventures were ever considered direct live feeds again. Space programs ensure controlled release of any space communication, image, and video footage. Space travel has become the world's most top secret.

Common sense tells us to realize, that this difficult process is for a reason. Wondering if we are not supposed to be in space or are, we? Convincing ourselves that we need to venture out into the stars for future survival. At some point something is coming or happening. What don't we know? What are they keeping from us?

Who new participating in democracy would lead to dissimulation? Trying to understand how a ballot box turned anyone into a participating citizen. Understanding today that very same ballot box carries submissions based on false representation. These individuals are restricted by a party influenced for higher government process. Providing those same individuals who were once on a ballot a source of security. Relinquishing their initial cause out of fear for losing income, better known as a seat. Governmental officials have always had an influential impact wherever money is. Even now, there is a new military branch called Space Force.

We are given an opportunity to partake with our own removal of existence. Depletion by which party? We pay tax to build and sustain government. We are not even a part of our own regulation. We are workers for our own government business, a dispensable write off. You won't be called on endless you're not meeting your quota, or your file is in discrepancy. Your file is that card they give you and it assigns a number to you. Known as a social security card. We don't have tickets, when a disaster comes, and it will. You will see how fast we become obedient disposable workers. You are not supposed to know this type of information. One thing is certain, the examples from the past have become the present. All tying into space travel because it carries the heftiest financial cost to accomplish.

Space agencies constant introduction of meaningless information seems all to purposely done to please or disinterest the public. New information with substance is a gold mine these days. Time travel is finding itself in direct correlation with space travel. Deception is being deployed, for what reason is the question? Are we really being protected from ourselves? At what point does it become an unbelievable excuse? Space travel has become unrealistic notion, dubbed for only belonging to big screen movie theater scripts.

Selective and manipulated educational perspective has superseded the average household's intellectual concept of space related topics. The careless release of information is issued to the public in a sense of complete disregard. As if the subject of space or space travel has become simple speculation, unrealistic theory, and stands completely oblivious to the population today. Until reusable remote-controlled rocket boosters fly across the Los Angeles sky, maybe then will notify the public, but it's been happening since back in two thousand and fifteen. We heard of only one instance because of the mass chaotic response of the public and there has been over hundreds of these test launches. At the completion of the test launch schedule, with success of course this technology allows for a refurbished implementation into the process of space travel. Possibly reducing the outlandish cost of the process? Unless it is consumed entirely by our government or military at which time this technology will undoubtably disappear under another classified file.

Space travel restrictions have become a negative epidemic to any notion of space curiosity. A subject that has gone obsolete can be our demise as a civilization. Space travel has become the human unknown due to the dark deception of government suppression. Perhaps reference a young child's fear of water at an early age. Children's inexperience of functionality within water can lead to a greater hurdle to overcome in the future of the being. Searching for a bigger insightful hypothesis, think of the child before even attempting to enter a body of water for the first time. It must be accomplished with great sensitivity. Earths human and humanoids negative perspective about space might eliminate the notion, of them ever wanting to entertain the idea of leaving Earth. Sooner than later this might not be an option. Looking ahead what if that's the whole point. Never needing to relocate over seventy five percent of Earths human or humanoids after a natural or preventable disaster that fails. Could very well be our fate by design?

It would be an assumption, but terraforming a planet has a very small window for error and to say the least we are moving fast. Abruptly fast, in comparison to previous outer space accomplishments. Remember this will probably be the biggest world challenge, civilization has ever known. Currently we are in the process of terraforming Mars. Specific applicants with wide ranging skills have been expected to become the first human to inhabit the neighboring planet.

While being accompanied by a small team of astronauts with previous experience and minimal military force with newer advanced weapons customized to the red planet. Majority of the population might prefer to remain here on Earth but to be honest most won't have a choice. The danger of space travel is always a priority, but the cost is ultimately the main reason.

The method of manipulation behind space travel is not so coincidental. Looming among us could be the demise of our own existence. My research concludes that life was inhabited first on our dormant neighboring planet Mars. My theory is clear enough to suggest that at the wake of another planet and amongst the same universe only one planet can be active at one time. At which time after the wake of Earth did Mars reach its dormant state?

Unknown atmosphere requirements. Leading my perspective to believe that Earth is the paradox of space. The spaces vacuum brings molecules, sub-atomic atoms, quarks, leptons, and particles that are used to counteract with Earths effects. Earth's magnetic poles and gases create a magnetism effect when collided together with molecules, sub-atomic atoms, quarks, leptons, and particles from outer space. The reaction while colliding with our atmospheric gases create a total of a three-hundred-mile-thick field around our planet. Only ten miles out of three hundred of our atmospheres are considered to really be the primary field of energy shielding Earth. My hypothesis indicates a new concept integrated with the old perspective of our Earth's atmosphere. Officially indicating a magnetism effect, in association with Earth's magnetic poles and gases these collide with space molecules, sub-atomic atoms, quarks, leptons, and particles from outer space. Suggesting our Earth's atmosphere is made of a (quantum system of magnetic gas.) This system, state of linear function, combination, of quantum uncertainty does not exist amongst physics today, but is my work of the future.

Current Atmosphere: $N+O+H_2O$

Actual Atmosphere - Quantum System of Magnetic Gas: $Fe+Co+Ni+Ar+N+O+7H_2O(g)$

$$N+I+M+SZ\ldots?$$

    Official research goes as far as the funding allows. Funding is restricted by government and its associations as direct beneficiaries. The study amongst school, scholar's, and further educational desire's on specific subject are off limits. Work done in sensitive areas are controlled and threatened by higher superseding officials. New discoveries have become negative impacts amongst traditional study on a wide range of educational practices in fear of losing institution and contract funding. As it is even easier now to be removed, replaced, or transferred out of projects. Thankfully we have independent and private scholars. Only negative notion of leaving advanced concepts to theory, but as we know from the advanced studies of the past we are leading in the future.

    Monitoring, regulating, and implementing a controlled system of education starts at a very young age out of government design. On a traditional sense, we don't see any issue with say a report card for an example. What about the curriculum? Other countries today have students participating amongst student exchange programs at such a young age. As high school level primary course work. By the time they return to their countries they will have graduated with working experience degrees of choice at eighteen years of age. Travel, function, and receive income adjustment in coordination to their accredited trainee, title, or position.

American educational structure consists of an exaggeration of general educational knowledge that eventually does not transition to their position of actual employment. Further education is beyond frustrating for students of all ages. Immediately after spending an additional four years in high school. Most students eventually dropped out. After this became a problem, they dropped the mandated test scores to ensure positive graduation rate. They could not have people dropping out and failing to complete the process as well. The systems design is more to discourage further education and development. While promoting the working class into the traditional liaison by design?

Since we are hurdling through space and time presently our future suffers by the waste side. We conquer space travel but can't figure out an efficient system of any kind for ourselves amongst our own planet. Not out of coincidence. Financial debt looms over our nation, our military force has become a part of the traditional norm compared to the rest of world. As we stand united today our country is betting the house on planets of rare elements. Among the stars there is a possibility that some planets bared rare elements on a large scale. The implementation of our new military branch Space Force confirms the urgency to be first at all cost. Launching us into the battle of planets or WWIII. This future power struggle can only be won from the stars. A battle of two power houses among Earth would leave nothing left to rule over today.

James M. Gonzales / The Codex
Answer Key Of The Ancient Astronaut Theory

The space race would seem far from over. Asking the question of the universe? Help me see what they cannot. We are only instrumenting the music, it's really played from above. A new strike of oil? Not exactly, but funding pours in. Just imagine a planet with large patches of gold crusted all over the surface and underneath to the crust. Meteorites crash all over the earth, but this search in space might be a costly hoax. As most meteorites are magnetic? Which would indicate how they get pulled out of space and directed toward earth to begin with. Remember the quantum system of magnetic gas. Our atmosphere, and magnetic poles. The evolution of our own existence consumes our fate but still I have faith we will prevail. Space travel is established and even if it's not our savior this page is testament of perseverance. We won't stop working until we find whatever the answer may be for our civilization.

CHAPTER THREE

The Worlds Ignored

What if you have your own interaction with another worldly entity. Maybe just one you cannot explain to be a part of the traditional daily normal functions. Fear of disbelief our falling on deaf ears. Convince yourself otherwise, segregation of your own thoughts, and think you're just exaggerating? Move on is what we are almost programmed to do. In most cases regarding extraterrestrial, unexplainable, godlike and tragic events. By coincidence or design?

Evidence is in every direction documented in detail within history, archeology, and religion publicly. We attach the world's best to report on the matter to bring clarity of how special these details from the past share a different story from what is portrayed amongst our civilization. Even then it's not enough. We put admission prices on tickets for things like museums, menageries, and nonprofit organizations.

Turning pieces of ancient history, archeology, and religion into a system of business. Decreasing the principle value of these subjects from a psychological perspective. Most never realize putting these categories together will give you the key to our creation. Negative results trigger a repetitive circulation. We all seen it before people showing up to church donating abruptly moving on with their day nothing stuck during or after mass. Continuing contributions to the government cycle one way or another. A double edge sword crippling and taxing our ability to see outside the working order of expectational demand. The greatest monumental pyramid, hieroglyphic depictions, unexplainable humanoid skeleton remains, or the new and old testament of the good book are not seen to be actual direction only an optional resource of information.

Are we searching for the truth of our existence or are we just going through the motions of reality? The perception of time limits our opinion to reason and matter. Foolhardy faith in illusion of organization doesn't mean guaranteed assurance from chaos. Past, present, and future notions of time indicate our demise by thy own hand. Stretching for the stars as done in the past. We are finding new ways to destroy our existence. Not preventative methods to protect our evolution. Instead of educating truth we promote working order around basic education.

Keeping space at the forefront of a race is an illusion of perception. Terraforming another planet is more of an ultimatum of our existence. Only so many can make the trip and the remaining will suffer the same fate as planet earth. In the meantime, our preparation consists of concealment on a massive scale. Inheriting, hiding, or destroying any and all signs of life on Mars.

Earth is the main transportation terminus between space and subsidiary neighboring planets. This explains all the strange sightings in our sky. Most cannot see because when you look up its only to take a deep breath but to others, they see the stars amongst them. An appalling feeling lingers when your all alone amongst the night sky and the stars shine their brightest. Suddenly you really are not alone. Changing your perspective. An elevated heart rate leaving implicational feelings of a center stage perhaps. This doesn't happen often at all because most citizens must report for employment at first light. Putting them at rest through open night skies.

Extended air traffic control reports flood the emergency lines. As usual if it's not a real emergency your directed to the non-emergency line. Is an unidentified air craft alluding to a crisis? Maybe not among today's society. Remembering we are already among the stars brings new perception. The concept of unidentified has been diluted with doubt that it could just be us.

Numerous reports reference a wide range of crafts being spotted. What about the ones that are malfunctioning or have crash landed? Concerned citizens call it in. Twenty minutes later the military unit arrives to confiscate or recover? The callers get an estranged visit from the men in black referring to themselves as government officials. Shortly then after those same callers have trouble referencing that incident until years later. Not going to jump the gun here but maybe this is the real evidence to ensure two things. One it's in participation with our government. Two it is our own creation or developed technology from the past.

Professional mental health practitioners receive new case studies. Working with patients they cross examine hundreds of times. Hypnotize and even have some who are willingly to undergo lie detection and pass with flying colors. All while being questioned about abduction situation by unexplainable extraterrestrial entities. Easily dismissed by the community because this style of evidence is not admissible in a traditional scenario of questioning. What about our governments known affiliation with this unknown entity? If this extraterrestrial entity is really a degenerated version of our human beings from the past with more advanced knowledge. Are they treating and manipulating our DNA with an advanced medical procedure? In almost every abduction case there is some memory of being operated on by these humanistic like beings in an extraterrestrial craft. Then they are simply released by the abductors.

For far too long the government has separated themselves from tragedy and chaos. The truth is of voluntary interaction and selfish gain. Certain grey areas keep the government comfortable within their regulation of governing. Confidentiality amongst different military levels of top security is just the minimal standard of classified. During President Clinton campaign his curiosity was on every major news network asking to know about the world biggest kept secret.

It would not seem too farfetched to suggest that our government is securely protected. Even against itself. The President of the United States is off limits to our relation or affiliations with extraterrestrial entities amongst our own planet. That is because in order to address these beings it must be done from within a world leader platform, not national or head of states. As these beings are spread amongst the united nation not as hostages or captives but as resourceful insurance. Making no difference to them because they have the edge in advanced intelligence and communication. In comparison to these different entities we are out matched. Leaving the question open if they are just cooperating for an unknown bigger purpose but what or why?

The world's top intelligence departments including scientists, physicists, and engineers have come into interactions with more advanced technology. Most instances are very public. It's always denied by government. It's a little hard to use a crazed discharged cover story on decorated physicist with top level security clearance. These are instances that the public is shocked by but are trained to move on from. Then in turn never demand answers for. Answers beyond just the traditional questions. What about the question that would and will ultimately determine our fate as a civilization down the road for the long haul? How come we have possession of these beings and their crafts but supposedly can't figure out the crafts capabilities. This is above a reverse engineering crisis.

These are issue of world security. The beings we have captured for whatever reason throughout time have multiplied amongst our custody. Collaboration and housing. Ensuring underground secured facilities for these beings. Thus, has been established amongst our bases along the western region of the United States.

At what point do we realize this is not a collaboration but a reform of the world's direction. Right underneath us? We have received no new intelligence or aid of information. We are swimming in the dark with no way to interact with these beings. If that's really the case who is Jrod, Jiren, and Valiant? All known by these names to be extraterrestrial entities that have been amongst our planet. Are we just providing our information and allowing access to our capabilities? Now more than ever it would seem so and without any doubt.

After shuffling information across to the major university laboratories around the United States we concluded a few new elements. We watched our entire military implode. Baring witness of nuclear test launches in space simply dissipated by the worlds supposed unknown.

Retired national security personnel review tape with head line news live across our country. Reiterating the concept that this type of occurrence is taking place all over our nation on major and small scales. Not just on random instance but all the time. To the point that when it is reported even to our own military forces there is really no sense of urgency within the overseeing government officials. The fear and dangers are only within our own selves. Knowing all life matters and is not to be played with for we have seen the effects of war.

What about the pilot who was trying to engage the unidentified air craft that was televised and discussed with they retired military security director? This incident is a real live scenario that took place. The pilot did in fact respond accordingly as he was willing to put his life on the line for our national security. He was not notified that this interaction is to be they expectational standard by any means. Since when is the military or civilian life secondary? In the United States of America every life matter's or does it anymore?

All while we harbor a handful of craft that we cannot operate. These very same crafts constantly break into our air space. Toy around with our most advanced fighter jets on national television. While many assume these being are visitors from another galaxy or these beings are in collaboration with our government.

Making light sport with potential enemy technology or any interaction with our military forces with a negative outcome reflects poorly on our nation. While these interactions might seem harmless now, but the level of danger presented to our own cadre in pursuit is a high-risk situation that should not be taking place. Even worse our countries security is at risk when we cannot protect all its citizens? This just might be too comfortable for safe keeping. Especially when military and civilian lives are at risk to the unknown.

While neighboring nation leaders make off air verbal communication mistakes while being recorded. Headline news sky rockets putting the world on notice about the collaboration with these beings. These extraterrestrial beings are not just giving us air fight or flight trouble. The problem goes deeper under the water than we have ever been or can go. These extraterrestrial beings would seem to have mastered the concept of time travel. For we have been superseded.

Elaborating on what we have all been seeing. As humanoids that evolved from humans with the aid of extraterrestrial beings. It's all just us. Everything from the extraterrestrial entities that are known to earth come from us. Hovering craft in and out of our waters, unknown traveling crafts, breaking light speed, and the laws of gravity. Accepting that all these new technological advancements that seem unrealistic have been around for hundreds of thousands of years. We are simply learning to use and create our own. In collaboration with past and future ancestral beings.

Huge deforming health risk from being in space and outer space travel for prolonged periods of time causes human deterioration. Putting stress on the body by activating and manipulating humanoid characteristics. Utilizing different or unknown genetic DNA markers. Transforming human beings. Just one tour to the moon and back won't complete this type of transformation. For you to be considered an extraterrestrial entity you will need to have been in space for years. Multiplying your age by hundreds and thousands of years in reference to Earths time.

Based on Earths time cycle of twenty-four-hour periods etcetera. In comparison of earth to space time. The further you get away from gravity time begins to move faster. Keep in mind there is little to no detectable gravity in space. Time among space differentiates dramatically from Earth. Remaining in space for a prolonged period will complete the time travel cycle. Considering we as a civilization have already accomplished time travel. Today we simply are trying to manipulate it at our own discretion. This becoming the pinnacle of the world most ignored.

CHAPTER FOUR

Gods of Man or Extraterrestrial?

The root of man or extraterrestrial is one in the same. Today amongst theorist and scholars it is a common understanding that we as humans to earth are not essentially made from earth. A conceptual seeded notion of evolution. Actively we are the primary effect of earths ability to a human extent. In contribution to the evolutionary properties gifted from our planet. At some point we elevated from evolved humanoid primates. The minds curiosity settles in. Especially when we our wondering in search of understanding of oneself. Glancing at the youth creates more confusion. Seeing the vicious cycle amongst other mammals. What have we become?

Looking back for the answers. Everything is distorted in depiction. Strange languages few understand or claim to translate. How could there be so much disconnect from the past versions of our very own selves. We have evolved to the next generation of beings? This confusion of transition is extreme. Making sense of our own developmental process. Knowing that we are the slowest developing beings among the surface of our own planet. Raising no questions as we turned this concept into inspirational stories, movie, or television. Something of a miracle.

Man, of the past and extraterrestrial beings of the future. Just as all our predecessors before? Not exactly. Ancient civilizations stress the concept of being visited and given assistance. Across the globe. The Gods from above came from the stars. Only today we have become humanoids of extraterrestrial gods from the ancient past, present, and future time. Descendants of human beings

As the human current time frame consist of mixed illusion and manipulation of the working majority. The general population alludes in denial and continues to bear witness of the unexplainable paradox. Financial struggle looms a fault to maintain material existence keeps humanoids restrained by human traits. Separating evolution from accruing at its official rate of potential, altered we have been. Sparked, making no matter the system was recognized, adopted, and scorched. The operation of society has been decided for us today. The transition from a superior being has been lost. Backwards, at the root of laborer just above a slave on the totem pole. By our own, today we serve the newly appointed and governed system. On the current time line sacrifice of time for currency takes precedence. Our true God today is government for who we serve as mandated until the next world war.

The educational aspiration of Americans has lost its abundance among the United States population. Trending habits and beliefs carry amongst populations. Understanding the importance of your psychological impression can help your current civilizational perspective. Knowing what the valuable source of interest is keeps us in sync on all platform.

Currency fuels the void of validity. Work is giving our humanoid population a manipulated sense of fulfillment in the form of daily achievement. At which time it blocks the sense of our own ability or self-existence. Keeping us mesmerized, confused, and prone to mimicking what everybody else in the same vicinity is doing. Small or minimal achievement seem more like life long goals. The burden of never-ending production for an unknown cause. If you achieve enough it consumes your reality making you a tapped resource.

Going to employment, entertainment, and running errands for mandated responsibilities. This brief list concludes the working-class humanoid life cycle. Everything else is considered a self-inflicted problem that traditionally makes the original working cycle more difficult to maintain. Mixing the mandatory requirements in massive volume with human desire is the recipe for gluttony but for who to blame? Working bodies don't have enough time to conclude self sufficiency independently. We appointed to participate in an invisible governed system that makes our lives adjust to this lifeless cycle. Making us more like a device. Hints my reference to humanoid verses human. We are trying to completely rid of any human flaws left. It's not efficient enough for the mass production needs and requirements?

How could so much be nothing? Finding ourselves operating more and more like the devices used in our everyday lives that we cannot go without. Sold under anticipations of a piece of the future we help fund. Our commitment to efficiency or inherited habit has triggered this humanoid like characteristic within our supposed human functionality. We are under the assumption that we are accomplishing but ultimately have nothing. What are we supposed to do? Why are we here on this planet? The average can barely make rent yet alone will probably never own a home among this very piece of land they fund today. Go to the bank just to find out how much you don't qualify for a loan. It's no worries, because you're paying for that very same banks needs with no questions asked. Losing more valuable emotional attached relation to endure more of a work load. Completing whatever is left of the transformation. Human consumed by humanoid functions. A lost perception or actual reality of our civilization today. Look around what's so different between you and the unit next to you. Who has more than the next? Purchasing more materialistic items and trying to maintain a standard of living that's beyond traditional necessity. While others of our own suffer alluding initial responses, that's their fault? How can we be gods?

We appoint combination formulas, tracking code, or identification number to monitor people. A direct reference to a traditional working employee or human being. Humanoids converted humans with device like characteristics. Programmed to adapt to modern tradition and exploitation at its finest. Leaning the ideal normal standard to be of extraterrestrial order. Today, compared to ancient times of barter and trade shows humanity has truly lost its way amongst human civilization as we once known it to be. This is how far away we truly have evolved from human beings. Following a system of convenience powered with our own repetitious cycle to become rid of and obsolete from any remaining human characteristics.

Humanoid a true device of the governed. Developed throughout time by a system that preferably adopted its citizens as dependable contributing resourceful working aid. Rather than slavery, elite regime, and controversial dictatorship. I honorably introduce the United States of America, the land of the humbly gathered. Peace keeper of war. Derived on adjustments of the most extreme scenarios. In fear of the wildly unknown, polished by the dirtiest way, and the nurturer of those in need with a beneficial subsidy.

The precedence is a true grim reaper. A Purchaser of the soul looming at trade of an empty promise too freedom under the highest regulation. Under voluntary opinions for your rightful sacrifice to gain in turn your ballot to vote for a meaningless option for which you will accept tax for system participation. Now at which time you will inherit the benefit to work congratulations?

Interactive advanced resource of technology is the current standing of human classification. Destined by a repetitive system that consumes yourself and those you leave behind when your time is done. Work more, make less, and never have enough. Repeat repeatedly. Then this is how you realize what has been being explained to you. Not just by some outlandish author, but an actual or former lost participant amongst the same system? Are you listening or just hearing me now? Remember the repetitiveness, just before we start creating life outside of our natural reproduction process because it's always never enough. Select genetic markers and match making will be the next form of human reproduction. Super humans are of a more efficient way. The best and least flawed beings for a greater form of productivity.

Shell-shocked but to the government this is what we appointed them to do. This notion of advanced being did not just spring up on us by surprise. Civilizations has been under transformation since the beginning of our existence as mammals. The intrigue of curiosity to know at what exact point did we evolve beyond just being traditional human beings?

In ancient times there are highlighted scenarios and hieroglyphics that are depicted of multiple phenomenon. Suggesting advanced civilizations back in ancient times from to the temples and pyramids found around the world. Evolving beyond man is our newest but oldest accomplishment. Halting the civilizations direction. Lost humans of extraterrestrial design.

Grounding our interaction and functional capabilities. Prioritizing the emphasis of production, enterprise, and government. Freedom to a contribution limit of interest. The initial concept at the turn of the century was short lived. Reaching this point didn't really live up to the hype. In theory we should be space suited and booted. In substitution for the good old shirt, jeans, and sneakers. Flying or hovering craft should have replaced the busted auto industry. Sky scraping, floating cities, and wireless everything to replace the run-down metro areas.

Far too much of an investment for temporary residents? A decision of god like control by an evolved humanoid race of government with extraterrestrial capabilities? Not you and me. We are just humans. The governed population is most of the working force that supplies funding, labor, system resource, and is the primary analytical test run of disposable beings. Rest assured they are happily doing so. It's the people like you and me. The ones that march to the beat of their own drum. In search of purpose with a desire for more explanation, reason, and accountability. That understand the madness and foresee the foolhardiness. This selfish fugazi consumes us all. Soon it will be illegal to think for yourself or outside the box.

Who knew the concept of a ballot box was going to be the exclusion of our freedom? Although throughout time hope is steadfast even as we see more regulation and decreased options of choice. Knowing we have given the title, deed, and trust away under false pretense. Our sacrifice under essence of a greater cause of the future we build. Selfish notions of the future while time escaped our effort. For the next generation our children. The traditional culture normal trend leads most of the motivational assumption of purpose.

Man, of extraterrestrial beings in search for the power of gods. Human evolution to what extent? We have transitioned to a new standard of existence. The physical representation of modern human characteristics cannot be denied. Indication of genetic manipulation would seem to be confirmed. This is where the answers should be clarified amongst society today. The perception of the working class is of a weakened state from its repetitive nature. Hope for the civilization's identity is lost within the work force we have become.

Can we say we are a seeded civilization or one that is in the process of being seeded? It would seem to be the first option with less and less abduction activity. Transforming the last of the original human entities. Leaving the last of the mammal elements to rid from our process. The remaining action of previous associating functions have also been altered and manipulated. We are presently in a new humanoid era. Reiterating the concept of our essence as interactive advanced technological resources.

The greatest human application in civilization has been a technological device. Not by coincidence. Definitively due to the turn of the century? Not quite, after years of waiting our perception relinquishes hope of a new advanced futuristic life style. Today almost every human has access to enough technology to go beyond our planet in their pocket. God like amongst today's mammals. Extraterrestrial humanoids evolving from humans of the planet earth.

CHAPTER FIVE

Ancient Histories Gods from Above

Ever since the beginning of our civilization these questions have been asked. Are we the only ones? What other forms are like us, none? Is there life among the other stars? Looking back at ancient history for general perspective about human life. Findings, the root of confusion. At some point we must accept that the information that's being discovered tells a different story about the past civilizations of humanity.

You and I compared to each other can easily be considered quite different. In association to one humanoid from current time today. In comparison to a humanoid even as little as three hundred years ago would undoubtably result in unexplainable difference. Not only within the genetic DNA markers but can you imagine the difference from a living, functioning, or communicating perspective. This exact example can be used to correlate the mental or psychological belief system in refence amongst today's paradox.

Opening our mind to the reality of an advanced technological perspective. That even today we are just scratching the surface to. First, we must get past the bias and non-bias psychological notion of criticism that hinders our minds vision. Due to the governed traditional perspective that manipulates life on an individual scale today.

Changing or adjusting your perspective is like switching the radio channel inside your motor vehicle. Considerably the same topics but referenced from a different point of view. Elaborated notion of the same song to a different beat. Remember the multibody dynamic perspective theory (MBDPT). You must keep your informational knowledge as a resourceful tool. It's never to be used for segregated measures. Blurry mental depth stems from this problem it infects us all. Penetrating, the strongest bonds. Bring light to the deepest issues or personal divides among our civilization. Everyone has an opinion that differs from the next person. We are finding things that change everyone's perspective. Leaving only one definitive assumption? Truth of the past referencing the gods from above.

We have not been ready for what is coming. Our inherited system of life has changed the outlook of so many. Twisted, micromanaged, conflict, of interests. What about the undoubtably righteous? The for certain. It becomes information but remember information must be received. The ancient history text depicts an entirely different perception than what is being concluded from a test tube. Keep in mind information is the key. Both research and discovery are relative facts based on opinion. Interest has become the scale of due process.

Us, me, you, and I are focused for self in search of more of what ever they tell us? Who tells us? The life you live for. Whatever that maybe? We are divided amongst each other. Regardless of residing within a nation of peace. We are separated and only choose to congregate at the sign of mutual benefit. Ancient histories Gods from above were known for weighing potential on a whole population basis. Is it safe to say that we may have already committed the underlining treachery by abusing our existence? As done in the Triassic and Jurassic era.

In the entire system of your life you sacrifice to gain supposed progress. Beyond our needs. Regardless of informational awareness of problems from the past world inhabitants. It's still selfish under foolhardy perspective completed with an illusion of dominance and power. Here and now. Knowledge is key. Work for more. I want, not need. I get and need more of nothing but it's a reason to keep going? How can we receive greater knowledge if we cannot manage what we have already been given? Ancient histories gods from above all around the world practiced and preached about repression. Temperance would be the fruit of the divine a gift of the merciful to be god like.

Ancient history tells us we co-existed with some of the best species from each respected era. Today we live in a collaborated existence. If this is a race, we got a head start and are still losing. Further and further we are sliding backwards. Each day we spend in a misled direction of life. Options of material resource have become a habitual standard for today's freedom. Human existence revolves around a shopping mall, hand gun, and a dreadful job. The greatest abilities we have are unknown to ourselves because of alienated roots of our own existence. Ancient history of the gods from above has been dismissed by our evolution today.

Today these gods from above are alluded or referenced to by a church. Majority of the civilization or many even then don't attend. Assuming they are too busy working for self-first.

This is our way of living life in the twenty first century. Even then on average most claim religion but its unpracticed. Worship is praised to Allah or a God's son as our lord and savior Jesus Christ. Keeping in mind there is a vast range of different religions.

I am just giving my readers a glimpse of they particularly lager followings for educational comparison and depth. Again, there is absolutely nothing wrong with being religious. I am myself. To me ancient astronaut theory is not a religion or movement but a very respectable information resource.

The ancient gods from above compared amongst todays traditional perspective still leads us to the same conclusion. That we are a product of a more advanced beings. At some point and time throughout history we have been manipulated to accelerate our traditional or original genetic DNA makeup. This perspective is firm in the sense that we have accomplished the time travel paradox. My scientific method: multibody dynamic perspective theory (MBDPT) is important because it elaborates a clear understanding by utilizing the deriving categories to establish this point of view. Maybe not every being from outer space or that comes into our planet are particularly of direct relation.

Undoubtably so, we're among, and are they extraterrestrial presence. We predominantly are responsible for the visible eye witness accounts or interaction that are reported within our own planet. Gifted a power to join the stars or become what long ago was considered of the gods from above. Amongst ancient time research, today we turn a blind eye if there is no personal gain of interest.

Most citizens find that heart felt moment. Coming across special or unique items of ancient history or historic monuments. The topic alone grabs almost everyone's attention. History's gods from above. Just as quick as the specific moment intrigues your mind its over. That very thought of how special something can be is quickly gone. Snapped back to our actual reality of life today commonly known or referenced to as your working calendar.

See its just that very concept that controls your minds perception. Way before you can get to the point or develop too much of your natural opinion. This governed systemic life cycle will pull you back into a looming reality of nothing. An overbearing work load, materialistic items, and a whirlwind of debt.

We learn to accept it, some hate it, others deal with it, and just think things are interesting. Try to adjust to a partner or significant others perspective. Some may relate. Others just because it makes some sense and never come into your own thoughts. Whatever the reason your consumed by the system. On that note only one perspective matters there is no need to hold your self-hostage. Especially because of a typical perspective. Dare yourself to be different. Embrace your own mind set and you might just shock yourself.

The stars flood our sky's we can't help but look up and wonder. Is it fear? Are we afraid? If that's the case, why? Because of change. This governed process has enabled us to become dependent upon it. There is no demand for supply. We are self-efficient beings. Ancient gods from above wasn't a slogan to the indigenous population of humanoids across our globe. They were visited. We have been seeded throughout time.

There is no explanation for you and I. We were not born this way. Look back and check our history. Before Adam and Eve in the good book. Descending genetic markers trace are genetics back to Neanderthals. Meaning were a part of the primate evolutionary tree. Along with monkeys, apes, humans, etc. There is up to five different types of humanoid skeletons that have been recovered from the ground we walk above. Representing actual evidence that we have evolved at an unexplainable rate. How could something so special be monotonous information to our current civilization.

Ancient gods from above blessed the united, humbled, gathered, righteous, and repressed. For we have sacrificed enough to hold testament and become godly ourselves. Now we have unfortunately lost sight of what has been given and do not remember it can easily be taken away. As has been done in the past. Our success is in the eyes of the beholden because of them we shall have hope. To live for a purpose not a moment. As we possess the ability to make a difference for a positive cause. Finally start to live and not just survive. Adjust and receive the grace of your creator. This is the greatest power to achieve for it is selfless and its path leads to peace. After all we are descendants of ancient gods from above.

THE END OF PART I

www.ingramcontent.com/pod-product-compliance
Lightning Source LLC
Chambersburg PA
CBHW040413220526
45473CB00004B/1226